ENERGY SECTOR STANDARD OF THE PEOPLE'S REPUBLIC OF CHINA

中华人民共和国能源行业标准

Code for Defining Land Requisition Treatment Scope of Hydropower Projects

水电工程建设征地处理范围界定规范

NB/T 10338-2019

Replace DL/T 5376-2007

Chief Development Department: China Renewable Energy Engineering Institute

Approval Department: National Energy Administration of the People's Republic of China

Implementation Date: July 1, 2020

China Water & Power Press

中国水利水电出版社

Beijing 2024

All rights reserved. No part of this publication may be reproduced, stored in a retrieval system, or transmitted in any form or by any means—electronic, mechanical, photocopying, recording or otherwise, without prior written permission of the publisher.

图书在版编目（CIP）数据

水电工程建设征地处理范围界定规范 : NB/T 10338
-2019 代替DL/T 5376-2007 = Code for Defining Land
Requisition Treatment Scope of Hydropower Projects
（NB/T 10338-2019 Replace DL/T 5376-2007） : 英文 /
国家能源局发布. -- 北京 : 中国水利水电出版社,
2024. 10. -- ISBN 978-7-5226-2777-9
Ⅰ. TV752-65
中国国家版本馆CIP数据核字第2024PZ6644号

ENERGY SECTOR STANDARD
OF THE PEOPLE'S REPUBLIC OF CHINA
中华人民共和国能源行业标准

Code for Defining Land Requisition Treatment Scope
of Hydropower Projects
水电工程建设征地处理范围界定规范
NB/T 10338-2019
Replace DL/T 5376-2007
（英文版）

Issued by National Energy Administration of the People's Republic of China
国家能源局　发布
Translation organized by China Renewable Energy Engineering Institute
水电水利规划设计总院　组织翻译
Published by China Water & Power Press
中国水利水电出版社　出版发行
　　Tel: (+ 86 10) 68545888　68545874
　　sales@mwr.gov.cn
　　Account name: China Water & Power Press
　　Address: No.1, Yuyuantan Nanlu, Haidian District, Beijing 100038, China
　　http://www.waterpub.com.cn
中国水利水电出版社微机排版中心　排版
北京中献拓方科技发展有限公司　印刷
184mm×260mm　16开本　1.75印张　55千字
2024年10月第1版　2024年10月第1次印刷
Price（定价）：￥280.00

Introduction

This English version is one of China's energy sector standard series in English. Its translation was organized by China Renewable Energy Engineering Institute authorized by National Energy Administration of the People's Republic of China in compliance with relevant procedures and stipulations. This English version was issued by National Energy Administration of the People's Republic of China in Announcement [2023] No. 1 dated February 6, 2023.

This version was translated from the Chinese Standard NB/T 10338-2019, *Code for Defining Land Requisition Treatment Scope of Hydropower Projects*, published by China Water & Power Press. The copyright is reserved by National Energy Administration of the People's Republic of China. In the event of any discrepancy in the implementation, the Chinese version shall prevail.

Many thanks go to the staff from the relevant standard development organizations and those who have provided generous assistance in the translation and review process.

For further improvement of the English version, any comments and suggestions are welcome and should be addressed to:

China Renewable Energy Engineering Institute
No. 2 Beixiaojie, Liupukang, Xicheng District, Beijing 100120, China
Website: www.creei.cn

Translating organizations:

China Renewable Energy Engineering Institute

POWERCHINA Chengdu Engineering Corporation Limited

Translating staff:

DAI Lei	LIU Wensheng	WANG Kui	SHI Xinchuan
WU Qifan	ZHANG Jiangping	LI Xiangfeng	LIU Yaxin
LIU Yingquan	XI Jinghua	ZHAI Hongguang	LIU Can

Review panel members:

GUO Wanzhen	China Renewable Energy Engineering Institute
LIU Xiaofen	POWERCHINA Zhongnan Engineering Corporation Limited
BIAN Bingqian	POWERCHINA Huadong Engineering Corporation Limited

QIE Chunsheng	Senior English Translator
JIA Haibo	POWERCHINA Kunming Engineering Corporation Limited
LIU Qing	POWERCHINA Northwest Engineering Corporation Limited
GUO Jie	POWERCHINA Beijing Engineering Corporation Limited
QI Wen	POWERCHINA Beijing Engineering Corporation Limited
LIU Hao	POWERCHINA Zhongnan Engineering Corporation Limited

National Energy Administration of the People's Republic of China

翻译出版说明

本译本为国家能源局委托水电水利规划设计总院按照有关程序和规定，统一组织翻译的能源行业标准英文版系列译本之一。2023年2月6日，国家能源局以2023年第1号公告予以公布。

本译本是根据中国水利水电出版社的《水电工程建设征地处理范围界定规范》NB/T 10338—2019翻译的，著作权归国家能源局所有。在使用过程中，如出现异议，以中文版为准。

本译本在翻译和审核过程中，本标准编制单位及编制组有关成员给予了积极协助。

为不断提高本译本的质量，欢迎使用者提出意见和建议，并反馈给水电水利规划设计总院。

地址：北京市西城区六铺炕北小街2号
邮编：100120
网址：www.creei.cn

本译本翻译单位：水电水利规划设计总院
中国电建集团成都勘测设计研究院有限公司

本译本翻译人员： 代 磊 刘文胜 汪 奎 石昕川
吴启凡 张江平 李湘峰 刘亚新
刘映泉 席景华 翟洪光 刘 灿

本译本审核人员：

郭万侦 水电水利规划设计总院

刘小芬 中国电建集团中南勘测设计研究院有限公司

卞炳乾 中国电建集团华东勘测设计研究院有限公司

郗春生 英语高级翻译

贾海波 中国电建集团昆明勘测设计研究院有限公司

柳 青 中国电建集团西北勘测设计研究院有限公司

郭 洁 中国电建集团北京勘测设计研究院有限公司

齐 文 中国电建集团北京勘测设计研究院有限公司

刘 昊 中国电建集团中南勘测设计研究院有限公司

国家能源局

Announcement of National Energy Administration of the People's Republic of China
[2019] No. 8

National Energy Administration of the People's Republic of China has approved and issued 152 energy sector standards including *Code for Operating and Overhauling of Excitation System of Small Hydropower Units* (Attachment 1), and the English version of 39 energy sector standards including *Code for Safe and Civilized Construction of Onshore Wind Power Projects* (Attachment 2).

Attachments: 1. Directory of Sector Standards
2. Directory of English Version of Sector Standards

National Energy Administration of the People's Republic of China

December 30, 2019

Attachment 1:

Directory of Sector Standards

Serial number	Standard No.	Title	Replaced standard No.	Adopted international standard No.	Approval date	Implementation date
...						
13	NB/T 10338-2019	Code for Defining Land Requisition Treatment Scope of Hydropower Projects	DL/T 5376-2007		2019-12-30	2020-07-01
...						

Foreword

According to the requirements of Document GNKJ [2015] No. 283 issued by National Energy Administration of the People's Republic of China, "Notice on Releasing the Development and Revision Plan of Energy Sector Standards in 2015", after extensive investigation and research, summarization of practical experience, and wide solicitation of opinions, the drafting group has prepared this code.

The main technical contents of this code include: reservoir-inundated area and impoundment-affected area, project construction area, boundary of land requisition and resettlement, boundary pillar arrangement design, and work requirements and outcomes of each stage.

The main technical contents revised are as follows:

— Adding the content of spare quarry and borrow areas.

— Adding the technical requirements for newly affected objects and their scoping in the operation and maintenance period.

— Deleting the content of the potentially impoundment-affected area.

— Deleting the content relating to the method for geologically identifying impoundment-affected areas such as landslide, bank collapse and immersion.

— Refining the content of design flood calculation, impoundment-affected area and project construction area.

— Refining the technical requirements for boundary of land requisition and resettlement.

— Refining the content of work requirements and outcomes of each stage.

National Energy Administration of the People's Republic of China is in charge of the administration of this code. China Renewable Energy Engineering Institute has proposed this code and is responsible for its routine management. Energy Sector Standardization Technical Committee on Hydropower Planning, Resettlement and Environmental Protection is responsible for the explanation of specific technical contents. Comments and suggestions in the implementation of this code should be addressed to:

China Renewable Energy Engineering Institute
No. 2 Beixiaojie, Liupukang, Xicheng District, Beijing 100120, China

Chief development organizations:

China Renewable Energy Engineering Institute

POWERCHINA Chengdu Engineering Corporation Limited

Participating development organizations:

POWERCHINA Beijing Engineering Corporation Limited

POWERCHINA Kunming Engineering Corporation Limited

Chief drafting staff:

LIU Huanyong	ZHANG Jiangping	LIU Yingquan	JIANG Xiehua
XI Jinghua	LI Xiangfeng	ZHAI Hongguang	XU Kaishou
CHEN Jing	XU Jing	DAI Lei	LIU Yuhan
LI Chongqing	JIANG Zhengliang	LI Hongyuan	ZHU Zhaocai
TANG Liangji	YU Lin	WU Liheng	WANG Xiaocong
YANG Changding	ZHANG Yong	JU Lin	

Review panel members:

GONG Heping	WANG Kui	ZHANG Yijun	GUO Wanzhen
YANG Dequan	WANG Chunyun	WANG Zhu'an	YAO Yingping
LIAO Wei	WANG Peng	LI Yongxin	YUWEN Zhenguo
WANG Zhengtao	YIN Zhongwu	ZHENG Shunxiang	HE Wei
WANG Xiaolan	BIAN Bingqian	ZHONG Guangyu	ZHAO Sheyi
CHUN Guangkui	LIU Yangwei	LI Bo	LI Tianhu
ZHANG Guoping	XIAN Enwei	PAN Shangxing	PIAO Ling
LI Shisheng			

Contents

1　General Provisions 1
2　Reservoir-Inundated Area and Impoundment-Affected Area 2
2.1　General Requirements 2
2.2　Reservoir-Inundated Area 2
2.3　Impoundment-Affected Area 6
3　Project Construction Area 8
4　Boundary of Land Requisition and Resettlement 9
5　Boundary Pillar Arrangement Design 10
6　Work Requirements and Outcomes of Each Stage 12
Explanation of Wording in This Code 14
List of Quoted Standards 15

1 General Provisions

1.0.1 This code is formulated with a view to standardizing the definition of land requisition treatment scope for hydropower projects.

1.0.2 This code is applicable to the definition of land requisition treatment scope for hydropower projects.

1.0.3 The land requisition treatment scope for hydropower projects shall include the reservoir-inundated area, impoundment-affected area, and project construction area. The land use for relocation and restoration works shall comply with the regulations of the state and provincial governments and relevant sector standards.

1.0.4 The definition of land requisition treatment scope for hydropower projects shall specify the design flood standard for each reservoir-inundated object; determine the scope of the land requisition and temporary land occupation for the reservoir-inundated area, impoundment-affected area and project construction area; plot the boundary of land requisition and resettlement; and conduct the boundary pillar arrangement design.

1.0.5 The definition of land requisition treatment scope for hydropower projects shall follow the principles below:

    1   The land use planning shall meet the project construction and operation needs while improving the land use efficiency.

    2   The impacts of project construction and operation on the surroundings should be minimized, and geological hazards should be avoided to ensure the land use safety.

    3   Less land use and less cultivated land occupation.

1.0.6 The definition of land requisition treatment scope for hydropower projects shall comply with the current sector standard DL/T 5064, *Specifications of Resettlement Planning and Designing for Hydroelectric Project*.

1.0.7 In addition to this code, the definition of land requisition treatment scope for hydropower projects shall comply with other current relevant standards of China.

2 Reservoir-Inundated Area and Impoundment-Affected Area

2.1 General Requirements

2.1.1 The reservoir-inundated area and impoundment-affected area shall include the areas submerged by the reservoir and the areas affected by the impoundment.

2.1.2 The reservoir-inundated area shall include the areas below the normal pool level, and the areas above the normal pool level but temporarily submerged by the reservoir flood backwater, wind wave, ship wave, ice jam, etc.

2.1.3 The impoundment-affected area shall include the areas of impoundment-induced landslide, bank collapse, immersion, bank deformation, waterlogging, and reservoir leakage, and other areas such as the surroundings and islands denied of fundamental production and living conditions.

2.2 Reservoir-Inundated Area

2.2.1 Reservoir-inundated area under the normal pool level shall start from the dam axis, and extend along the normal pool level to the intersection with the water surface profile of average annual flow in the natural river course.

2.2.2 For different reservoir-inundated objects in backwater area, their design flood standards shall differ. The backwater surface profiles for different periods shall be calculated to rationally determine their thinning-out sections of reservoir backwater.

2.2.3 The determination of the design flood standards for reservoir-inundated objects shall meet the following requirements:

1. The design flood standard for a reservoir-inundated object shall be determined according to its importance and inundation endurance, considering the reservoir regulating performance and operation mode, original flood control standard, safety, and economy.

2. The design flood standards for important reservoir-inundated objects, such as railways, highways, electricity facilities, telecommunication facilities, water conservancy facilities, and cultural and historic relics, shall comply with the current national standard GB 50201, *Standard for Flood Control*, and other relevant sector standards. If not specified in the standards, the design flood standard for a reservoir-inundated object may be determined according to its importance.

3 The design flood standards for reservoir-inundated objects should adopt the upper limit of design flood recurrence interval shown in Table 2.2.3. Otherwise, analysis and demonstration shall be made.

Table 2.2.3 Design flood standards for different reservoir-inundated objects

Reservoir-inundated object	Flood frequency (%)	Flood recurrence interval (a)
Cultivated land, garden plot	50 - 20	2 - 5
Forest land, grassland, unutilized land	Normal pool level	–
Rural residential area, town, general city, general industrial and mining area	10 - 5	10 - 20
Relatively important city, medium-sized industrial and mining area	5 - 2	20 - 50
Important city, important industrial and mining area	2 - 1	50 - 100

2.2.4 The calculation of the backwater surface profile of design flood shall comply with the current sector standard NB/T 35093, *Code for Calculation of Reservoir Backwater of Hydropower Projects*, and shall also meet the following requirements:

1 The backwater surface profiles in flood season and non-flood season shall be calculated respectively according to the reservoir operation mode.

2 For the project adopting staged impoundment, the backwater surface profiles shall be calculated by stage. For the project with a long construction duration and considerable increase in flood level due to project construction, the backwater surface profile during the construction period shall be calculated as needed.

3 If there is major tributary inflow or there are important reservoir-inundated objects in tributaries, the backwater surface profiles shall be calculated respectively for the following two combinations: the reservoir-inundating flood occurs in the tributary, and a given flood occurs in the main stream at the same time; the reservoir-inundating flood occurs in the main stream, and a given flood occurs in the tributary at the same time. The backwater surface profile shall be determined by the envelope of the combinations.

4 The calculation of backwater shall consider the effect of sediment deposition. The sediment deposition period shall be selected between 10 and 30 years, according to the sediment characteristics, reservoir operation mode, and importance of reservoir-inundated objects.

   5 The cross sections for backwater calculation shall be consistent with those for sediment scouring and deposition calculation. For the river reach with important reservoir-inundated objects or the thinning-out section of backwater, more cross sections shall be used as appropriate.

   6 The backwater surface profile of design flood shall be defined by the envelope formed by the flood backwater levels of the same frequency in different periods.

   7 The backwater surface profile should be calculated by the linear interpolation method if there is a great difference in the flood backwater level between two adjacent sections.

2.2.5 The thinning-out section of backwater under design flood and the end of the reservoir-inundated area shall be determined according to the following requirements:

   1 The thinning-out section used for the backwater calculation shall be the cross section where the water level difference between the backwater surface profile of the design flood and the natural water surface profile of the same frequency flood is 0.3 m.

   2 The end of the reservoir-inundated area shall be the point where the horizontal extension of the design flood level at the thinning-out section intersects with the average annual flow surface profile of the natural river course.

   3 If the design flood surface profile of cultivated land and garden plot is higher than the design flood surface profile of resident relocation in the upstream of the thinning-out section, the resident relocation area and the end of the reservoir-inundated area shall be defined by the design flood surface profile of cultivated land and garden plot.

2.2.6 The calculation of wave runup shall consider the factors such as wind speed and fetch. When the bank slope is below 45°, the fetch is within 30 km and the wind speed is below force 7 (14 m/s to 17 m/s), the wave runup should be calculated by the following empirical formulae:

$$h_p = 3.2Kh\tan\alpha \qquad (2.2.6\text{-}1)$$

$$h = 0.0208 V^{5/4} D^{1/3} \qquad (2.2.6\text{-}2)$$

where

  h_p is the wave runup (m);

  K is the coefficient relating to the roughness of the bank slope, which is taken as 0.77 to 1.00 for smooth and even artificial slope surfaces, such as block stone or concrete slab slope surfaces, and 0.5 to 0.7 for farmland ridges lower than 0.5 m;

  α is the bank slope (°);

  h is the height of wave in front of the bank slope (m);

  V is the speed of wind perpendicular to the bank slope (m/s);

  D is the fetch of wave on the windward side of the bank slope (km), which refers to the maximum straight line distance from the bank to the opposite bank.

2.2.7 The wave runup caused by ship travel shall be calculated according to the requirements of the shipping sector when there is a demand for navigation in the reservoir.

2.2.8 The reservoir-inundated area of ice-jam backwater shall be determined by the average pool level, average inflow and ice inflow during the ice surge period, taking into account the backwater surface profiles during freeze-up period and ice break-up period.

2.2.9 The determination of reservoir freeboard shall meet the following requirements:

  1 The reservoir freeboard shall be determined by the waves due to wind forces and ship travel, considering the impact of reservoir inundation on cultivated land and residential areas.

  2 Different reservoir-inundated objects may have different freeboards according to their importance.

  3 For the area immediately upstream of the dam, the wave runups due to wind forces and ship travel shall be calculated, respectively, and the larger value shall be taken as the reservoir freeboard. The reservoir freeboard for cultivated land shall be taken as 0.5 m when the calculated value is less than 0.5 m, and the reservoir freeboard for residential area shall be taken as 1.0 m when the calculated value is less than 1.0 m.

2.2.10 For a reservoir adopting staged impoundment, the reservoir-inundated area shall be determined by the backwater surface profiles of various reservoir-inundated objects calculated based on the project construction and impoundment plans, considering the water retaining condition of upstream cofferdam and the staged impoundment level. The extra reservoir-inundated area due to river closure shall not be included in the land requisition treatment scope.

2.2.11 The land heightened by engineering measure to regain its original use in the reservoir-inundated area shall be regarded as temporary land use.

2.2.12 Reservoir-inundated area shall consist of the area below the normal pool level, and the area defined by the envelope of temporarily reservoir-inundated area caused by reservoir flood backwater, water level raised by wind forces and ship wave, and ice-jam backwater.

2.3 Impoundment-Affected Area

2.3.1 The identification of impoundment-induced landslide, bank collapse, immersion, bank deformation, waterlogging, and reservoir leakage shall comply with the current sector standard NB/T 10129, *Specification for Preparation of Special Geological Report on Impoundment-Affected Area for Hydropower Projects*. The treatment scope shall be defined according to the importance of and hazards to the affected objects and based on the special geological report on impoundment-affected area of the hydropower project.

2.3.2 The importance of affected objects shall be classified in accordance with Table 2.3.2. The classification of hazards to the affected objects shall comply with the current sector standard NB/T 10129, *Specification for Preparation of Special Geological Report on Impoundment-Affected Area for Hydropower Projects*.

Table 2.3.2 Importance classification of affected objects

Importance	Affected object
Important	Buildings, important structures and important infrastructure
Less important	General infrastructure, cultivated land and garden plot

2.3.3 The area with reservoir bank landslide and deformation, where important affected objects or general infrastructure suffering from severe hazards exist, shall be identified as the treatment area of the objects or infrastructure.

2.3.4 The area of bank collapse where important or less important objects exist shall be identified as the treatment area of the objects.

2.3.5 The area of immersion, waterlogging and reservoir leakage, where important or less important objects exist and suffer from severe hazards, shall be identified as the treatment area of the objects.

2.3.6 The treatment area for the surroundings and islands denied of fundamental production and living conditions shall be determined through comprehensive analysis according to the cause and damage to the affected objects. The area shall be deemed as treatment area in any of the following cases:

1. The area where the production, living and shipping conditions are affected by the water diversion of the hydropower project.
2. The area around reservoir and island denied of fundamental production and living conditions caused by impoundment.

2.3.7 For a reservoir adopting staged impoundment, the impoundment-affected area should be defined according to Articles 2.3.1 to 2.3.6 of this code.

3 Project Construction Area

3.0.1 The project construction area shall include quarry and borrow areas, spoil areas, on-site access roads, construction facilities areas, camps and project management areas, which shall be classified by purpose into temporary and permanent land occupation areas.

3.0.2 The project construction area shall be defined according to the construction general layout, considering the land use purpose, land use conditions, and resettlement needs.

3.0.3 Permanent land occupation area shall include project structure areas, site access roads, permanent roads including the roads to the dam and powerhouse and between them, and roads in camps and project management area, as well as operational management camp site and ecological protection management area. Temporary land occupation area shall include quarry and borrow areas, spoil areas, construction facilities areas, and temporary roads for transporting spoils, exploiting construction materials, filling the dam, transporting concrete, and connecting construction facilities areas and contractor camps, as well as construction management and contractor camps. Spare quarry and borrow areas should not be included in the project construction area.

3.0.4 The overlap between project construction area and reservoir-inundated area shall be considered as part of reservoir-inundated area, while practically, considering the timeline of the project construction, this part may be requisitioned along with the rest of project construction area.

4 Boundary of Land Requisition and Resettlement

4.0.1 The boundary of land requisition and resettlement shall include the boundary of resident relocation and the boundary of land requisition, and shall be determined by the reservoir-inundated area, impoundment-affected area and project construction area.

4.0.2 The boundary of resident relocation shall be determined by house demolition needs in reservoir-inundated and impoundment-affected areas and project construction areas.

4.0.3 The land boundary shall include the boundary of land requisition, the boundary of temporary land occupation, and the boundary of land treatment, and shall meet the following requirements:

1. The boundary of land requisition shall enclose the land in reservoir-inundated area, and permanent land requisition in project construction area.

2. The boundary of temporary land occupation shall enclose elevated protected land in reservoir-inundated area, and temporary land occupation in project construction area.

3. The boundary of land treatment shall enclose the treatment-required areas beyond the land requisition boundary and the temporary land occupation boundary.

5 Boundary Pillar Arrangement Design

5.0.1 The resident relocation area and land requisition area shall be demarcated with boundary pillars. Boundary pillars may be classified into temporary and permanent boundary pillars.

5.0.2 The temporary boundary pillars and markers shall be set as required by inventory survey. Pillars may be made of wood, and markers may be made on tree trunks, rocks, walls, etc.

5.0.3 The permanent boundary pillars may be classified into main pillars and split-spaced pillars, and may be classified by material type into reinforced concrete pillars, carved pillars, steel pipe pillars and pool level signs. The permanent boundary pillars shall meet the following requirements:

1. The reinforced concrete boundary pillars should be prism. The main pillars should be 10 cm × 10 cm at the top, 20 cm × 20 cm at the bottom, and have a height of 100 cm to 140 cm. The split-spaced pillar should be 5 cm × 5 cm at the top, 10 cm × 10 cm at the bottom, and have a height of 50 cm to 70 cm.

2. The carved boundary pillars shall be engraved on the immovable rock, and the size of the groove should be 40 cm × 8 cm.

3. The steel pipe boundary pillars made of concrete and steel pipes should be 10 cm × 10 cm at the top, and have a height of 100 cm to 140 cm.

4. The pool level signs should have a size of 5 m × 3 m or 3 m × 2 m.

5.0.4 The layout of permanent boundary pillars shall meet the following requirements:

1. The control points for land requisition boundary should be provided with main boundary pillars, serving as the control pillars.

2. The spacing shall be 500 m to 1000 m between main boundary pillars, and 50 m to 200 m between split-spaced pillars. The adjacent pillars shall be visible to each other. A boundary pillar shall be set at each turning point.

3. The permanent boundary pillar shall be fixed by concreting or plain soil compacting. The permanent main pillar should be 15 cm to 25 cm above the ground, and the split-spaced pillar should be 10 cm to 15 cm above the ground. The allowable error of the set elevation is 10 cm for permanent main pillars and 30 cm for split-spaced pillars.

4. Permanent boundary pillars should adopt reinforced concrete pillars,

carved pillars, and steel pipe pillars. In addition, pool level signs shall be set near the sensitive areas such as cities, towns and important special items.

5.0.5 The numbering of boundary pillars and markers shall meet the following requirements:

1 Temporary boundary pillars and markers may be numbered temporarily.

2 The number of reinforced concrete boundary pillars and steel pipe boundary pillars shall be engraved on the top of the pillars. The carved boundary pillars shall be numbered and marked with the elevation underlined. The reinforced concrete boundary pillars, steel pipe boundary pillars and carved boundary pillars shall be numbered in a unified manner. The pool level signs shall be marked with water level symbol and characters.

3 Main and split-spaced pillars shall be separately numbered with letters and figures to indicate the boundary, treatment, affected object, main stream/tributary, and river bank.

5.0.6 The technical requirements for boundary pillar survey shall comply with the current sector standard NB/T 35029, *Code for Engineering Survey of Hydropower Projects*.

6 Work Requirements and Outcomes of Each Stage

6.0.1 The reservoir-inundated area, impoundment-affected area and project construction area shall be preliminarily proposed at the pre-feasibility study stage. The reservoir-inundated area generally includes the inundated areas of flat pool area and backwater. For the inundated area of backwater, the flood backwater surface profile may be calculated based on the preliminary normal pool level and design flood standards of different reservoir-inundated objects. The inundated area of flat pool area may be determined by considering a certain freeboard above the normal pool level. The impoundment-affected area may be proposed according to the preliminary conclusions of the engineering geological evaluation of the reservoir area and other influencing factors. The project construction area should be proposed based on the preliminary construction general layout. The reservoir-inundated area and impoundment-affected area of staged impoundment need not be considered. The sketch maps of reservoir-inundated area, impoundment-affected area and project construction area shall be prepared at the pre-feasibility study stage.

6.0.2 The land requisition treatment scope shall be determined in the resettlement plan and report at the feasibility study stage. The main work and outcomes shall meet the following requirements:

1. The scopes of reservoir-inundated area, impoundment-affected area and project construction area shall be determined based on the special reports on normal pool level analysis, geology of the impoundment-affected area, and construction general layout design, considering the affected objects.

2. A chapter regarding the land requisition treatment scope shall be prepared.

3. The sketch maps of reservoir-inundated area, impoundment-affected area and project construction area shall be prepared.

4. A resettlement boundary map of reservoir-inundated area shall be prepared, indicating the resident relocation and land requisition boundaries at each impoundment stage and the backwater level and treatment boundary. A resettlement boundary map of impoundment-affected area shall be prepared, indicating the resettlement and land requisition boundaries and the coordinates of inflection points. A resettlement boundary map of project construction area shall be prepared, indicating the permanent and temporary land occupation

areas and the coordinates of inflection points.

5 The technical requirements for arranging temporary boundary pillars shall be proposed as required by inventory survey.

6.0.3 At the resettlement implementation stage, the permanent boundary pillar arrangement shall be conducted, and the survey and design report and layout shall be provided. If there is any change, the land requisition treatment scope shall be reviewed, and the land requisition treatment scope map and resettlement boundary map shall be adjusted accordingly. For the cultivated land and garden plot affected by impoundment-induced landslide and bank deformation, the treatment scope shall be determined according to the actual damage.

6.0.4 At the operation and maintenance stage, for the newly added impoundment-affected area, the objects and scope to be treated shall be determined according to the actual impacts and technical requirements for defining the impoundment-affected area.

Explanation of Wording in This Code

1 Words used for different degrees of strictness are explained as follows in order to mark the differences in executing the requirements in this code.

    1) Words denoting a very strict or mandatory requirement:

    "Must" is used for affirmation; "must not" for negation.

    2) Words denoting a strict requirement under normal conditions:

    "Shall" is used for affirmation; "shall not" for negation.

    3) Words denoting a permission of a slight choice or an indication of the most suitable choice when conditions permit:

    "Should" is used for affirmation; "should not" for negation.

    4) "May" is used to express the option available, sometimes with the conditional permit.

2 "Shall meet the requirements of…" or "shall comply with…" is used in this code to indicate that it is necessary to comply with the requirements stipulated in other relative standards and codes.

List of Quoted Standards

GB 50201, *Standard for Flood Control*

NB/T 10129, *Specification for Preparation of Special Geological Report on Impoundment-Affected Area for Hydropower Projects*

NB/T 35029, *Code for Engineering Survey of Hydropower Projects*

NB/T 35093, *Code for Calculation of Reservoir Backwater of Hydropower Projects*

DL/T 5064, *Specifications of Resettlement Planning and Designing for Hydroelectric Project*